To Dorie,

your mission is to visit as many of these World Heritage sites as possible during your travels!

M, H, O + A

Toilets

Nature's call has never been so beautifully answered

Ladies and gentlemen, welcome to the ultimate toilet book.

As any experienced traveller knows, you can tell a whole lot about a place by its bathrooms. Whatever you prefer to call them – lavatory, loo, bog, khasi, thunderbox, dunny, washroom or water closet – toilets are a (sometimes opaque, often wide-open) window into the secret soul of a destination.

It's not just how well they're looked after that's revealing, but where they are positioned and the way they've been conceptualised, designed and decorated. Toilets so often transcend their primary function of being a convenience to become a work of art in their own right, or to make a cultural statement about the priorities, traditions and values of the venues, locations and communities they serve.

The lavatory is a great leveller – everyone feels the call of nature, every day – but, like any common species, being ubiquitous doesn't make it uniform. Around the planet (and beyond it, see page 12) toilets have followed various evolutionary pathways to best suit their environment.

In these pages you'll find porcelain pews with fantastic views, audacious attention-seeking urban outhouses, and eco-thrones made from sticks and stones in all sorts of wild settings, from precipitous mountain peaks to dusty deserts. So, wherever you're reading this, we hope you're sitting comfortably.

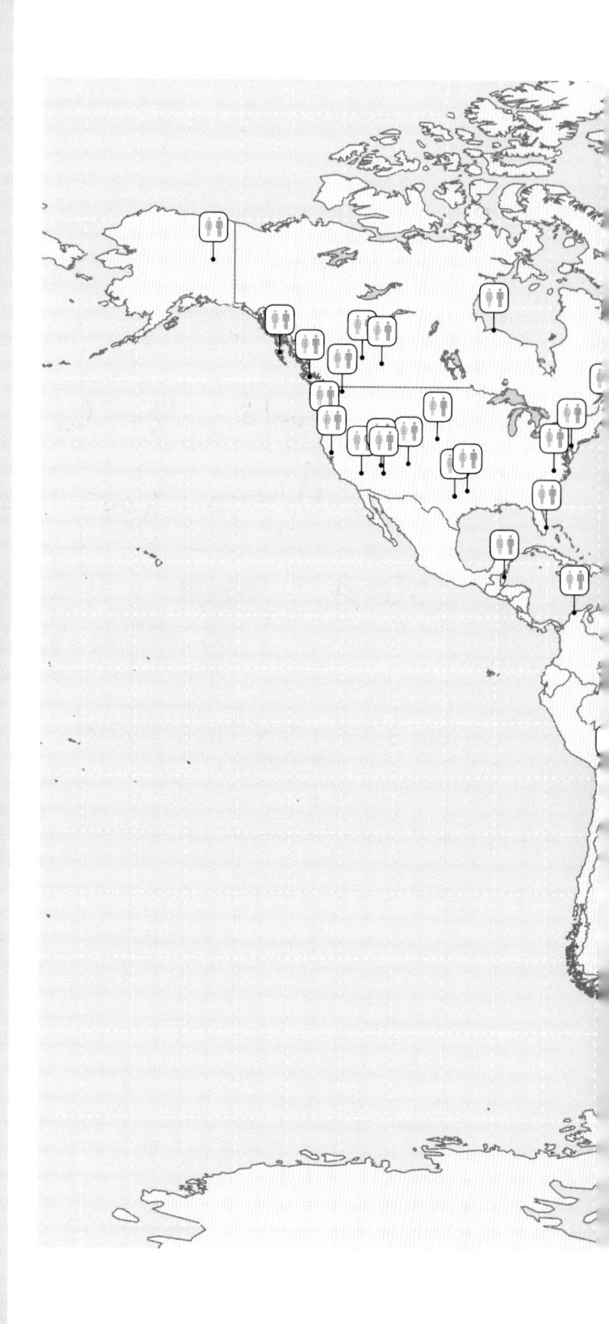

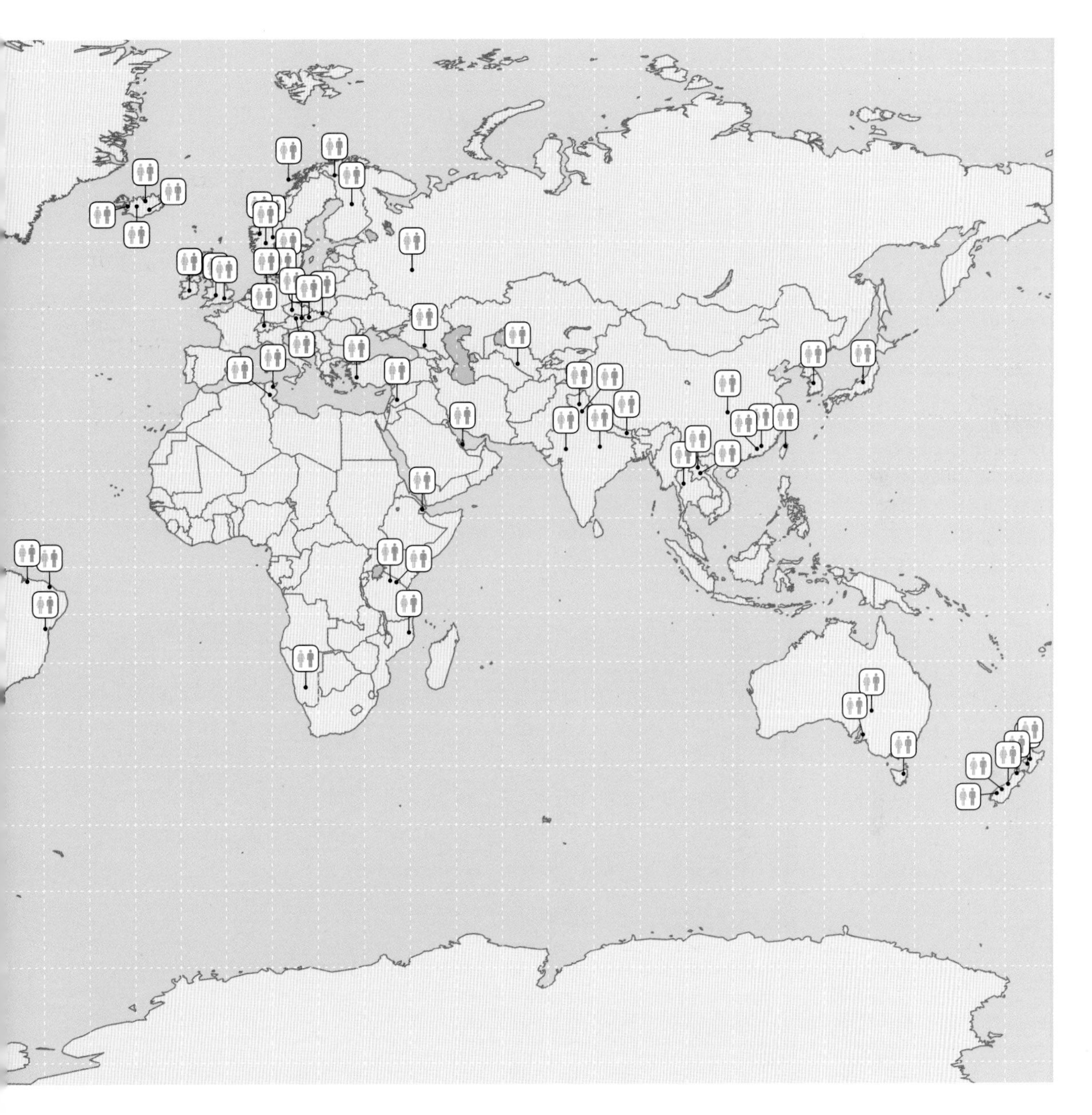

Lobster loos, Wellington, New Zealand

Spend a penny? Wellington, the capital of New Zealand, spent NZ$375,000 on architect Bret Thurston's boggly-eyed design for the public lavatories on the city's windswept waterfront. It is hoped that the two tentacles, armoured in orange steel, will attract tourists to Wellington, though it's a long way to go.

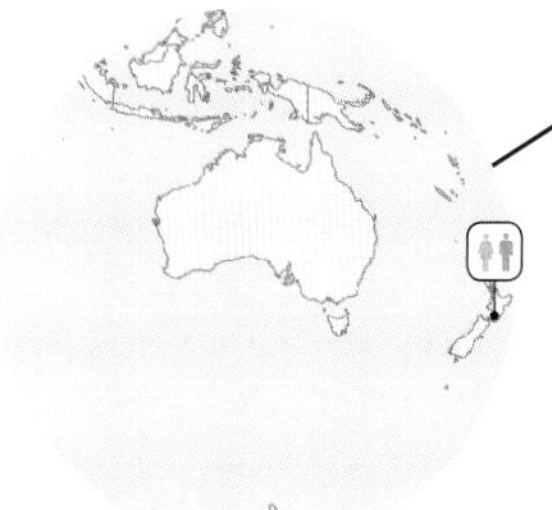

S 41° 17′ 5.5968″ E 174° 46′ 41.6604″

LINDA MCKIE / GETTY IMAGES ©

Alcatraz, San Francisco, USA

If you were a guard at Alcatraz high-security prison, you had to have a head for heights, even during your toilet breaks. Using these precarious watchtowers won't have been relaxing, but at least you'd have stellar views of San Francisco from your penitential perch.

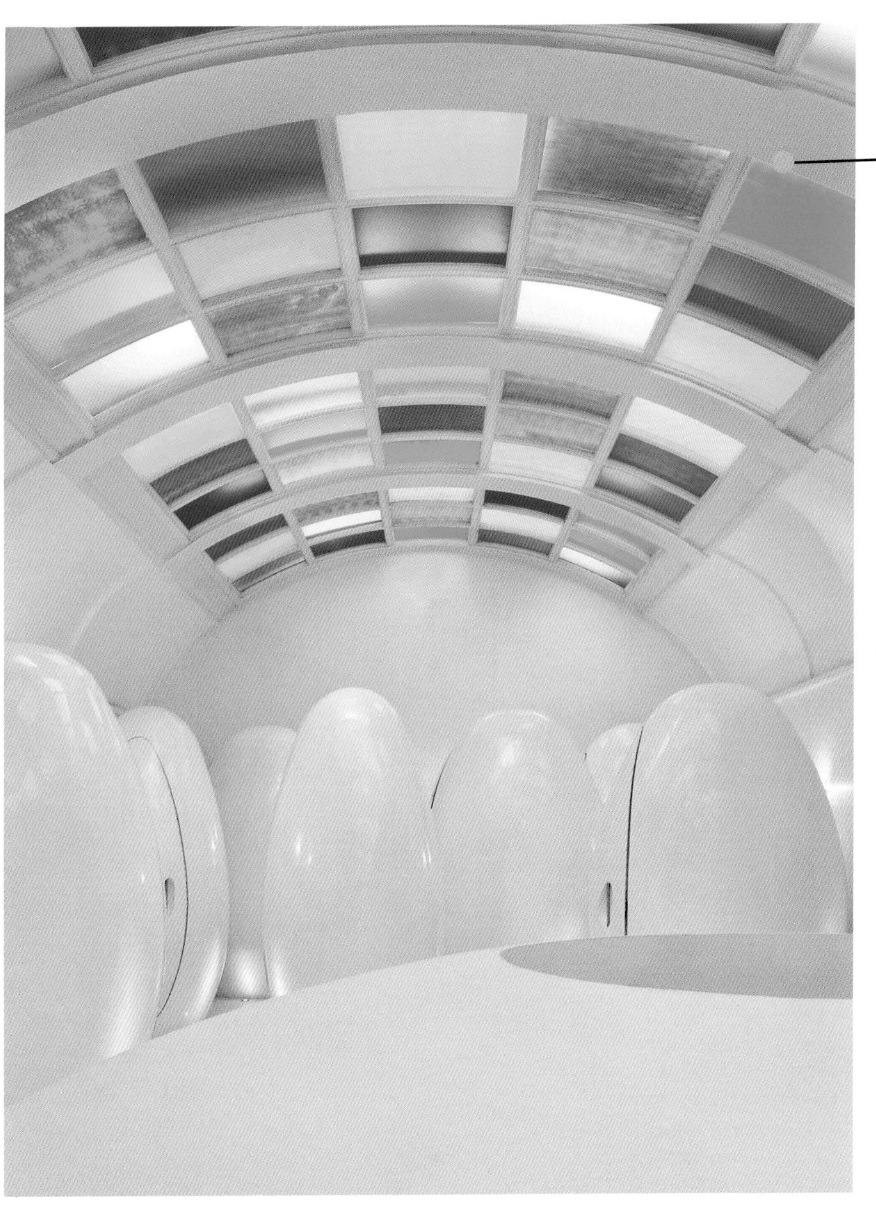

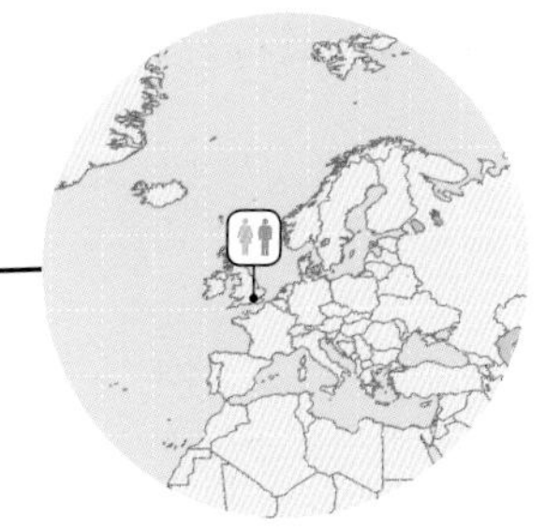

© ED REEVE

Sketch

London, UK

Don't panic! These aren't alien eggs waiting to hatch in a futuristic world. They're the famous toilet pods at Sketch restaurant, one of London's swankiest eateries. At least we think they are. As the night gets later, the lighting gets wilder.

Shard, London, UK

Enjoy a birds-eye view of London landmarks such as the Gherkin when you use the facilities in the Shard, the British capital's pointiest building, designed by architect Renzo Piano. The privilege of using the Shard's viewing platform will set you back £25.

Tundra toilet, Arctic Canada

The orange sail around this alfresco Arctic potty in Canada's far north is not there to protect anyone's modesty – it's there to prevent people's posteriors from being frozen solid to the seat in the -80°C (-112°F) temperatures that can grip the windchilled polar tundra. Few linger long enough to finish the crossword.

Bodie Ghost Town, California, USA

There might not be any riches remaining in the old gold-mining town preserved in Bodie State Historic Park but the restroom is a gem. With views over the badlands of California, northeast of Yosemite, this is one toilet trip you'll remember.

N 38° 12′ 50.9364″ W 119° 0′ 16.5636″

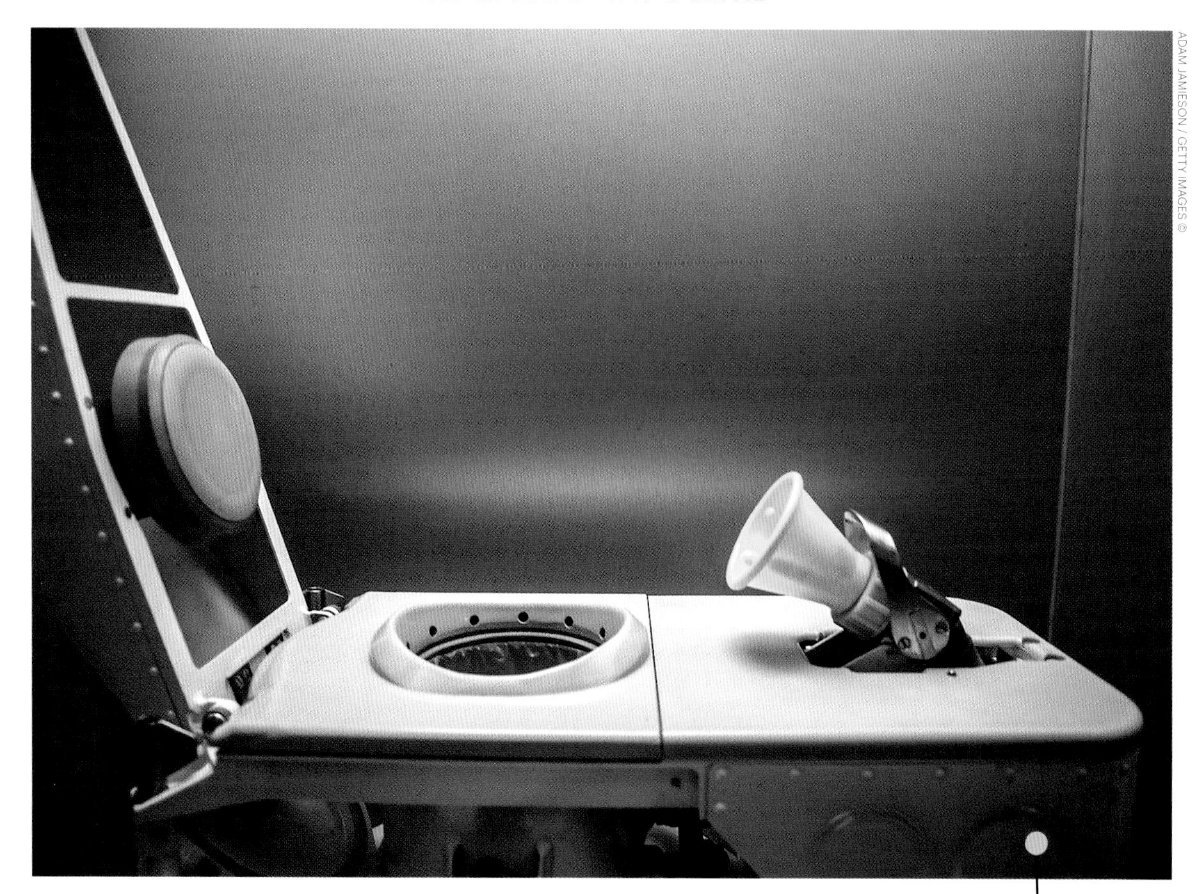

ADAM JAMIESON / GETTY IMAGES ©

Prototype space toilet

No, Earthling, this isn't an alien probing machine – it's a space toilet. Performing basic human functions in zero gravity is complicated. A suction system helps things travel in the desirable direction. Get it wrong, and you'll leave the ultimate floater – but at least, in space, no one can hear you scream.

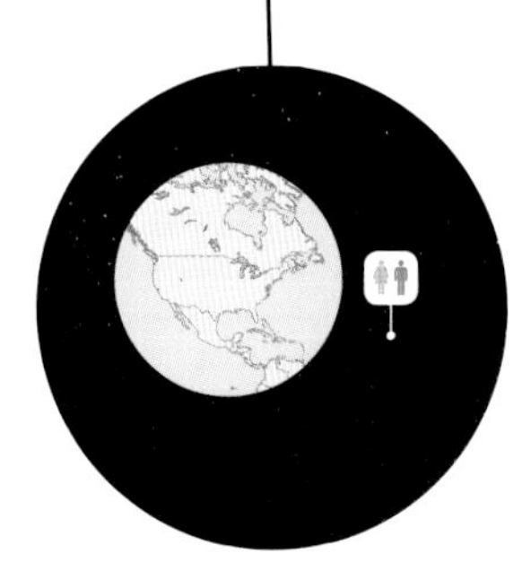

Jungle toilet, Vang Vieng, Laos

Vang Vieng in Laos was once infamous for raucous jungle parties, full of wasted Westerners tubing along the tree-lined Nam Song River. The illegal bars were closed in 2012, and now better-behaved travellers can enjoy a more tranquil experience – although answering a call of nature still feels pretty wild.

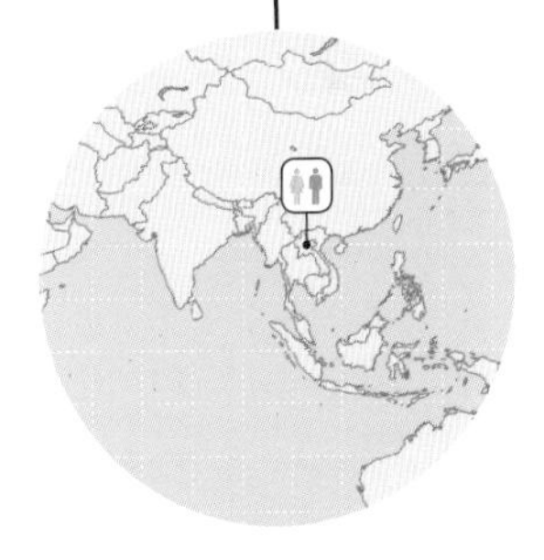

S 36° 21′ 8.9892″ E 174° 43′ 4.8792″

Public lavatories, Matakana, New Zealand

Locals in Matakana waited seven years and spent a pretty penny (NZ$400,000) to come face-to-face with their pouting public toilets, which provoked plaudits and protestation. Matakana lad Steffan de Haan's design is highly symbolic, from the facade to the ship-shape cubicles, a nod to the local boat-building industry.

S 36° 21′ 8.9892″ E 174° 43′ 4.8792″

N 51° 27′ 36.5292″ W 2° 28′ 30.0684″

Tardis, Warmley, UK

Sadly, this Tardis toilet, which materialised in a cafe garden beside the Bristol-to-Bath cycle path in 2014, is only a replica of the legendary laws-of-physics-defying police box that allows Dr Who to do his business, and you'd need more than a sonic screwdriver to get it moving. Inside the 'Who Loo', you'll find a flash, fully functional and flushable Victorian-style convenience, complete with sensor-operated flashing lights and sound effects. Strictly one time traveller at a time, although it's deceptively big inside...

POLICE PUBLIC CALL BOX

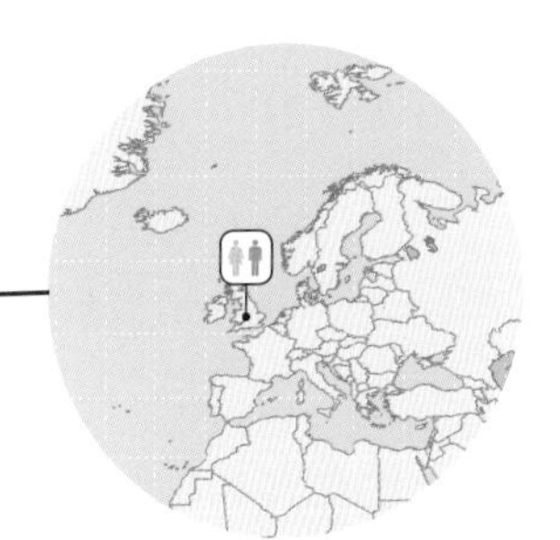

Simón Bolívar's washroom, Santa Marta, Colombia

Behind every great man there's a great... toilet. This was the last bathroom of revolutionary hero Simón Bolívar, who liberated most of South America. Bolívar was born in Venezuela, and lent his name to Bolivia, but he died in Quinta de San Pero Alejanrino, Colombia, where his washroom is preserved.

Public toilets, Moscow, Russia

This trio of bio toilets, standing sentinel by Ploshchad Revolyutsii metro station in the Red Square, is loudly and proudly emblazoned with intricate traditional red-and-gold Russian folk designs, and watched over by a matriarchal Muscovite who won't take any nonsense. Now wash your hands!

Forest outhouse, Pudasjärvi, Finland

There are no ventilation issues in this old outhouse, which clings on to a precarious existence atop a woody hill in Pudasjärvi, Northern Finland. Before the trees mounted a counter attack, the hill once hosted a fire lookout tower, and this tenacious toilet is the last structure standing.

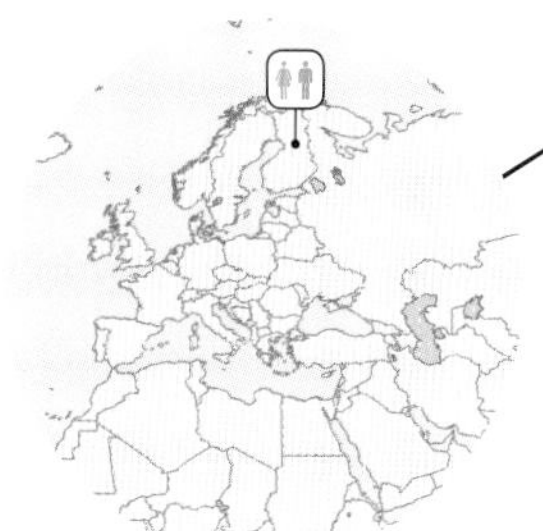

N 65° 21′ E 26° 59′

N 35° 41′ 37.824″ E 139° 42′ 12.777″

High-tech toilet, Tokyo, Japan

Putting the Wii into going for a wee, operating the toilet in the high-tech Tokyo suburb of Shinjuku is a little like negotiating the latest games console. Just be careful not to confuse the bidet button with the ejector seat.

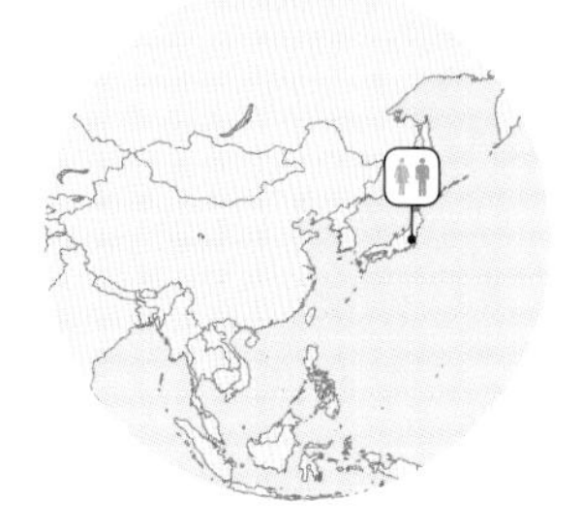

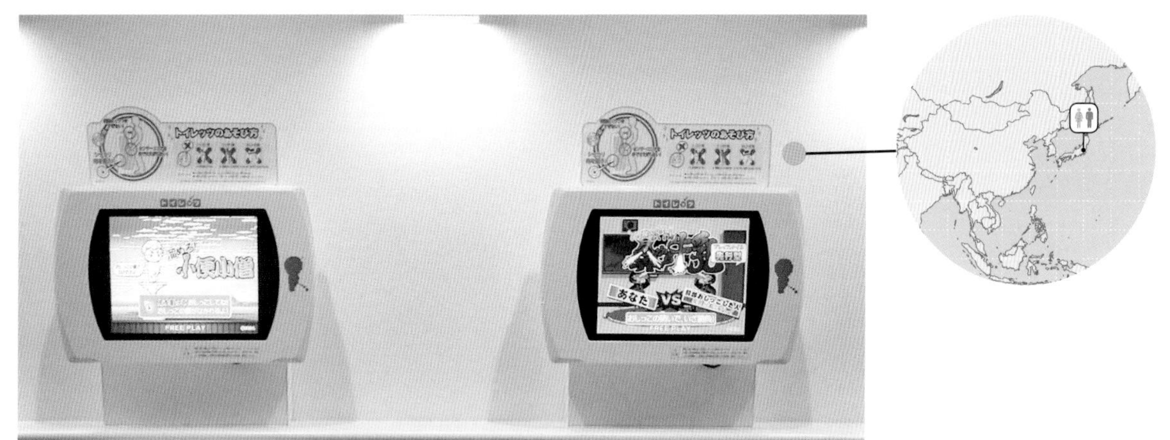

Toylet, Joypolis, Tokyo, Japan

There's never a dull moment in a Japanese restroom, not since Sega invented an interactive urinal system – otherwise known as the Toylet – where you can test your target skills or stream strength against your peeing peers. It's the 21st-century version of chasing a cigarette butt down a urinal.

Cobra toilet, Varanasi, India

When passing sacred water in the ladies' and gentlemen's street conveniences in the Ganges city of Varanasi, India, you can relax – Shiva's cobra has got your back. Shiva, one of the Hindu faith's triumvirate of gods, wears a cobra around his neck to signify power over dangerous creatures.

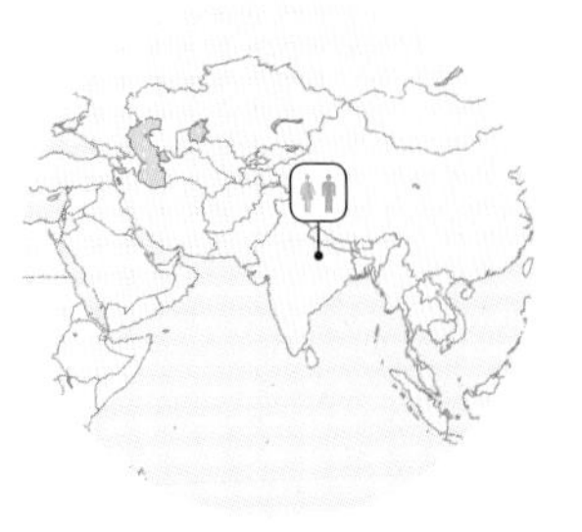

Outhouse, near Steamboat Springs, Colorado, USA

Any port in a storm – when you've got to go and it's 10 below, an outhouse in Colorado can provide a refuge, but you won't be lingering long on this far-from-hot seat. Temperatures can go well south of -20°C (-4°F) in the higher parts of the US state during winter.

Toilet in the desert, Jaisalmer, India

The Thar Desert – or Great Indian Desert – straddles the boundary of India and Pakistan. It's an incredibly arid area, but you can replenish its fluids at this lonely restroom near Jaisalmer, India, which even offers some protection from the infamous sandstorms that sometimes blast past.

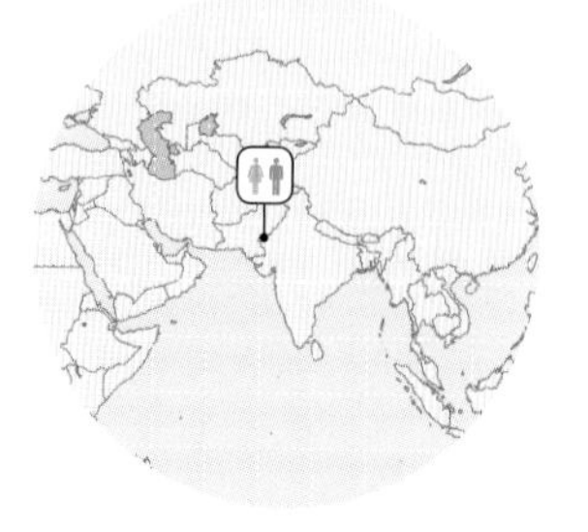

Desert restroom, Djibouti

Someone once dreamed of turning Arta Plage – a French military training area in the Djibouti desert – into some sort of Gulf of Aden oasis. It hasn't happened yet, but they did get as far as building toilets – even if you do need the courage of a commando to venture inside.

GENTS URINAL

Restrooms with a view, Chang La Pass, India

One of the unexpected effects of high altitude on the human body is the sudden need for a toilet stop – these strategically positioned public toilets perched on the top of Chang La, a 5425m (17,800ft)-high Himalaya pass near Leh in Ladakh, India, provide a restroom with a top-of-the-world view.

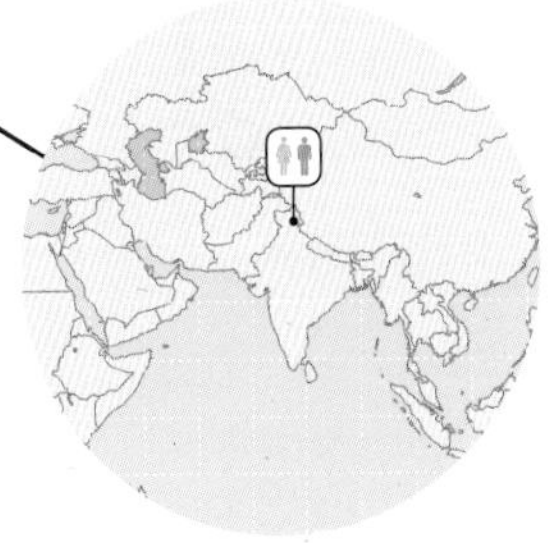

Etosha National Park, Namibia

There are many things you might hope to find in Namibia's sprawling 22,270-sq-km (8600-sq-mile) Etosha National Park – lions, leopards, giraffe, warthogs, even something super rare, such as the black rhino – but a well maintained public toilet... they're a long extinct species, surely? Apparently not.

KEN WELSH / DESIGN PICS / GETTY IMAGES ©

Thatched toilet, Gougane Barra, Ireland

Gougane Barra in County Cork, Ireland, has been a refuge – a retreat from the cruel world and a spot for quiet contemplation – since St Finbarr established an island monastery on the lake during the 6th century. These thatched toilets, tucked away in the forest park, continue that tradition.

Piano toilet, Chongqing, China

You'll never be caught short in Chongqing, China, which proudly holds the world record for the largest bathroom – a four-storey building boasting 1000 toilets. This colourful keyboard-shaped convenience is quite conservative in comparison to some of the city's more controversial creations, which include urinals in the shape of the Virgin Mary.

© IRA BERGER / STOCK PHOTO / ALAMY

Meatpacking district, New York City, USA

Manhattan's Meatpacking District has experienced an extraordinary evolution in the last half-century – the former home of the city's slaughterhouses saw some seedy subcultures sprout during the hedonistic 1980s, but now it's NYC's hippest 'hood, full of trendy clubs and bistros, like the Standard Grill, where you'll find this mind-bending bog.

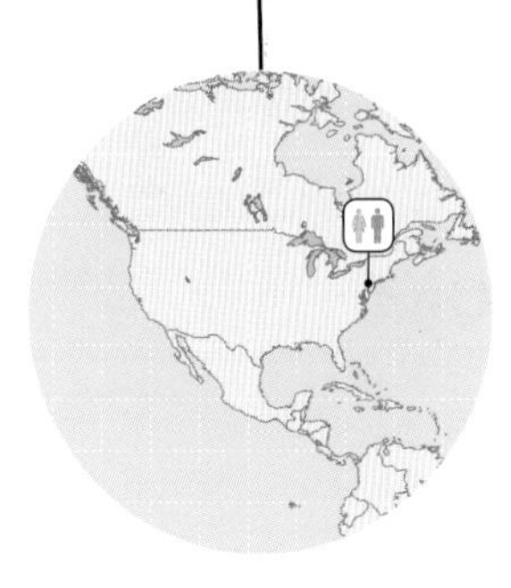

Trail restroom, Austin, Texas, USA

Hikers and bikers on the Lady Bird Lake Trail, which traces the banks of the Colorado River through Austin, Texas, can enjoy a gleaming state-of-the-art restroom, made from 49 steel plates strategically standing on end in a spiralling pattern to achieve maximum light and air flow, while guaranteeing some potty privacy.

N 30° 15′ 19.8540′′ 97° 44′ 23.2728′′ W

© PAUL FINKEL / PISTON DESIGN / MIRÓ RIVERA ARCHITECTS

TRAVELSTOCK44 / LOOK-FOTO / GETTY IMAGES ©

Peninsula Hotel, Hong Kong, China

Taking afternoon tea in the lobby of the Peninsula, Hong Kong's oldest hotel, might be classy, but to properly appreciate your surrounds, sneak off to the top-floor Felix Bar, where urinals positioned in front of a floor-to-ceiling glass panel allow you to pee and see cracking views of Victoria Harbour.

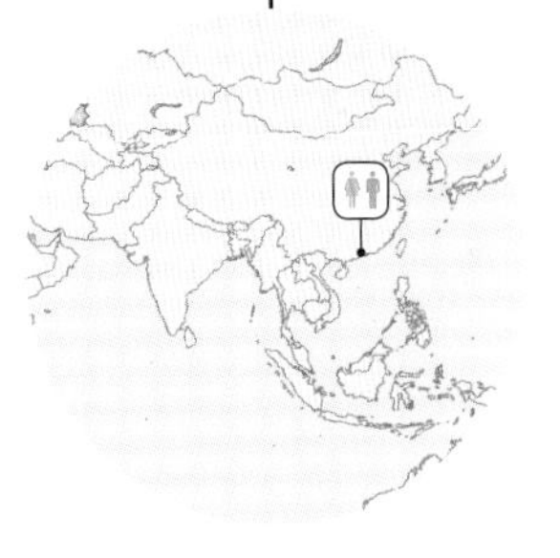

© JOHN MCINTIRE / 500PX

Winter toilet, Langidalur, Iceland

When Iceland's Eyjafjallajökull volcano last erupted in 2010, it filled the sky with smoke and emptied it of aircraft, with flights grounded all across Europe for days – an impressive effort, and something to contemplate, from your poo pew with a volcanic view in this campsite in Langidalur, Thorsmork.

S 3° 6′ 55.3824′′ E 37° 23′ 46.8816′′

Barafu Camp, Tanzania

Squatting on the edge of a cliff, 4600m up the flanks of Mt Kilimanjaro in Tanzania, the Barafu Camp khazi takes the concept of a long-drop toilet to an elevated level. *Pole pole* (slowly, slowly) is the standard mantra when climbing Africa's highest peak, but that adage doesn't apply here.

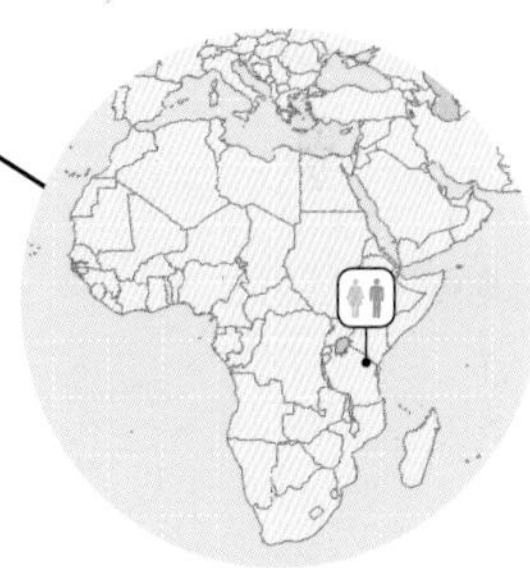

Monument Valley, Utah, USA

If you don't mind a mob of wild dogs standing guard as you attend to important business, then this outhouse in the surreal surrounds of Monument Valley, amid the wind-carved sand sculptures of the Colorado Plateau, is one for you. The Wild West valley is famous for its buttes, but some are best kept behind closed doors.

Krafla, Iceland

This ever-so-alfresco ablution station in the middle of the Icelandic outback, near Krafla Geothermal Power Station, is an enigma. No one seems to know who installed it, or why, but that doesn't worry happy hikers who, after stumbling across it, invariably Instagram images of themselves perched on the pan.

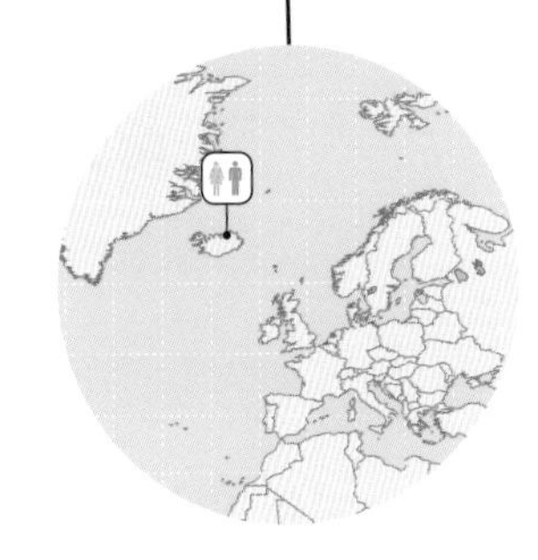

N 48° 12′ 34.0056″ E 16° 22′ 6.7512″

Adolf Loos public toilets, Vienna, Austria

Designed by the appropriately named Austrian architect Adolf Loos, the public toilets beneath street level along the famous Graben in Vienna, are a washroom wonder – although you'd never guess until you descend into the Art Nouveau netherland.

© HERBERT AIGNER / ALAMY

Built between 1909 and 1911, and complete with elegant dark wood and brass trimmings, magnificent mirrors and highly sympathetic mood lighting, these top toilets are a premier derrière parking place, if ever there was one.

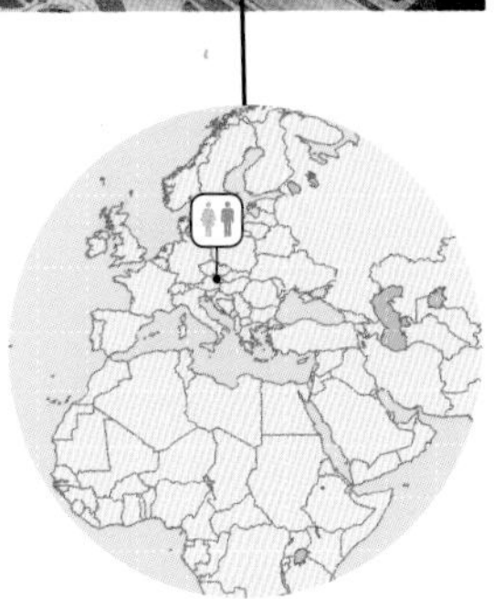

Museum of Islamic Art, Doha, Qatar

Although this structure (designed by Pritzker Prize-winning architect IM Pei) looks capable of blasting off into space and attaining warp speed within seconds, it actually has a slightly more prosaic purpose in life – as a public convenience in the park outside the Museum of Islamic Art in Doha.

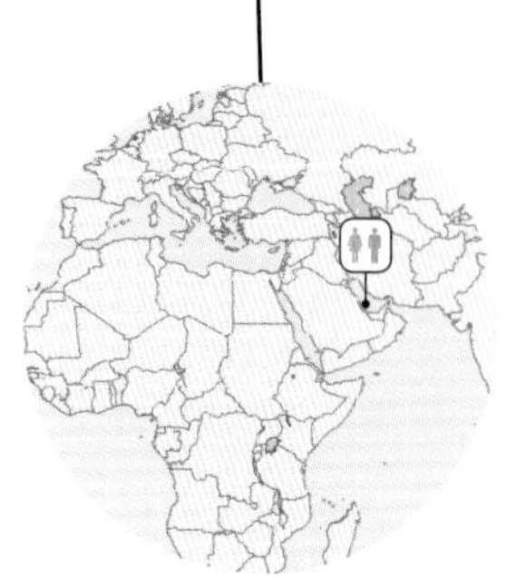

Penguin privy, Golden Bay, New Zealand

Colonies of little blue penguins have set up camp all around Golden Bay at the top of New Zealand's South Island, where the beaches are protected from the Pacific Ocean's tempestuous mood swings by the curve of Farewell Spit. Humans are apparently allowed to use these Pohara village toilets too.

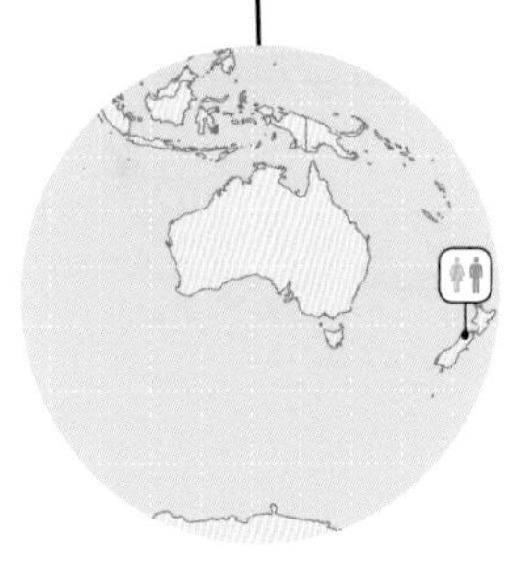

Fountain of toilets, Foshan, China

Made from 10,000 toilets, sinks and urinals, this fantastic flushing fountain graces Shiwan Park in Foshan, China, the world's ceramic capital. The installation, which is 100m (330ft) long and 5m (16ft) high, is the handy work of Chinese artist Shu Yong, who used factory seconds and pre-loved pans to create his masterpiece.

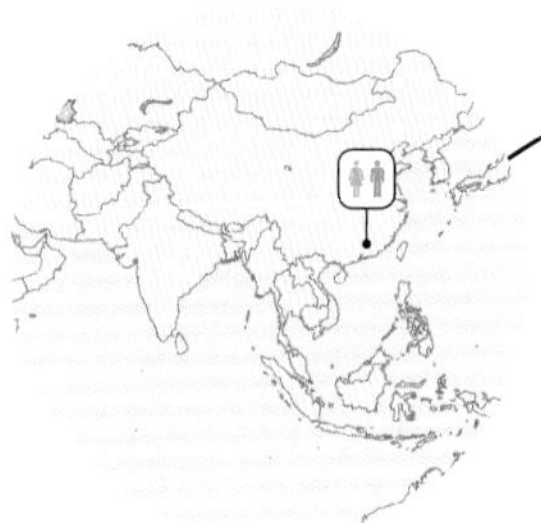

Scott Duncan Hut outhouse, Alberta, Canada

To reach Scott Duncan Hut, 19km (12 miles) from Lake Louise on the northwest ridge of Mt Daly in Canada's Banff National Park, you need to traverse the unforgiving terrain of the Wapta Icefields. The effort is well worth it, if only for an outhouse boasting the best view in the New World.

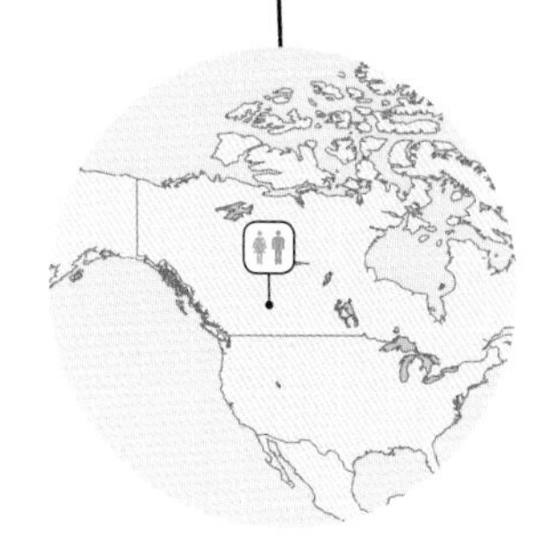

Long-drop, Olduvai Gorge, Tanzania

Tanzania's Olduvai Gorge is part of the Rift Valley, where the human species spent its formative years. Our ancient forebears went about their business here in the eastern Serengeti over 1.9 million years ago. Continuing in that tradition, this toilet features a seat directly overhanging the edge of the ravine.

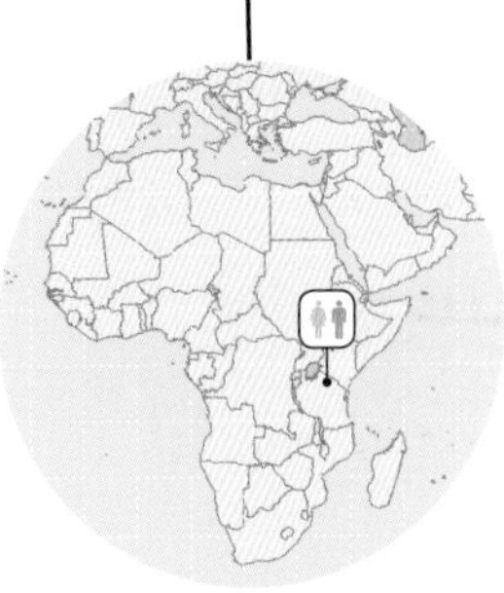

Alexanderplatz, Berlin, Germany

Berlin's Alexanderplatz – known to its friends as Alex – is a public square in the eastern part of Berlin, famous for the funky Weltzeituhr (a worldtime clock that shows the time across the globe), Hermann Henselmann's architecture, and its subterranean public toilets, with their futuristic facades.

Pop-up urinal, London, UK

We've had pop-up shops and bars, so it was only a matter of time before the pop-up potty raised its porcelain head. This hydraulic UriLift urinal can be summoned from the ground in Paddington, London, whenever it's required, and sent back to whence it came once business has been concluded.

N 27° 54′ 16.3764″ E 86° 52′ 16.5864″

Outhouse, Sagarmatha National Park, Nepal

At 6812m (22,350ft) high, eastern Nepal's Ama Dablam might not be one of the tallest Himalayan peaks, but it's possibly the most eye-poppingly beautiful. And there's no better place to sit and contemplate its magnificence than on the perfectly positioned throne at the lodge-village of Chukhung, 2000m (6560ft) beneath the summit.

N 27° 54′ 16.3764′′ E 86° 52′ 16.5864′′

Headlands Center for the Arts, San Francisco, USA

Housed in a series of artist-rehabilitated military buildings in historic Fort Barry, just north of San Francisco's Golden Gate Bridge, the Headlands Center for the Arts supports creative expression across a range of mediums – an ethos reflected even in Building 944's latrines, lovingly reimagined by artists Bruce Tomb and John Randolph.

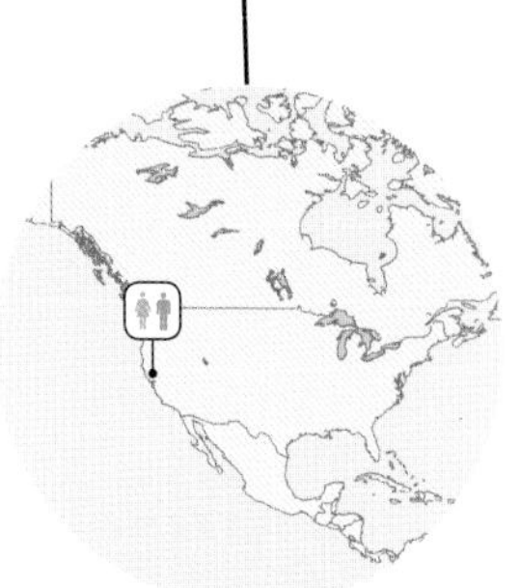

Patriotic privy, Oslo, Norway

Located on Oslo's Eidsvolls plass and designed by artist Lars Ramberg, this rabble-rousing trio of toilets sport the colours of Norway's flag – which it shares with France's Tricolore – and proclaim the three elements of the French revolutionary motto: *liberté* (freedom), *égalité* (equality) and *fraternité* (brotherhood). Inside, recordings of famous political speeches play.

N 34° 3′ 20.7432″ E 77° 40′ 1.2144″

Thiksey Monastry, Ladakh, India

Located on a hilltop about 20km (12.5 miles) from Leh in Ladakh, India, Thiksey is a Tibetan Buddhist monastery of the Yellow Hat (Gelugpa) sect. Lamas here live quiet lives dedicated to the cultivation of the view of emptiness... but at least they have a great valley vista from their 'panorama toilet'.

N 34° 3′ 20.7432″ E 77° 40′ 1.2144″

Kyzyl Kum desert, Uzbekistan

Covering parts of Uzbekistan, Kazakhstan, and Turkmenistan, the Kyzyl Kum is a 300,000-sq-km (116,000-sq-mile) sprawl of red sand, where temperatures are known to nudge 51°C (124°F). It's home to small, hardy agricultural communities and a big desert monitor that reaches 1.6m (5.2ft) long – one compelling reason to build toilets raised from the ground.

Mt Shuksan, Washington, USA

Facilities at Mt Shuksan's Sulphide Glacier base camp offer a grand vista of Mt Baker on a clear day, but they can be a bit breezy when the weather comes in. The peaks rise in North Cascades National Park in Whatcom County, Washington, just 19km (11.8 miles) shy of the Canadian border.

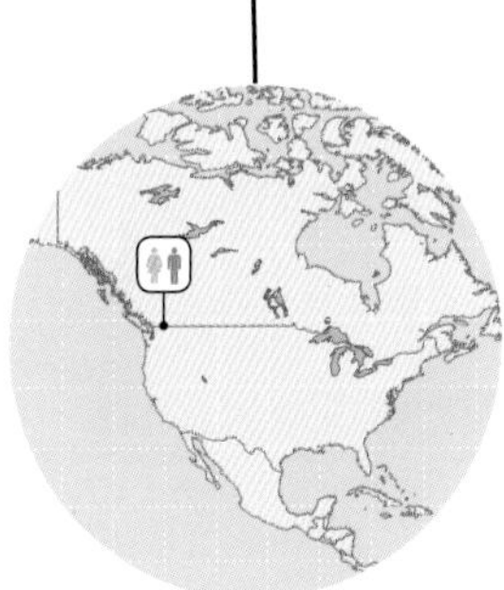

Public restroom, Kawakawa, New Zealand

Tourists typically leave the New Zealand town of Kawakawa talking about two things: the glow-worm caves in nearby Waiomio, and the endearingly idiosyncratic toilet block designed by Austrian artist Friedensreich Hundertwasser.

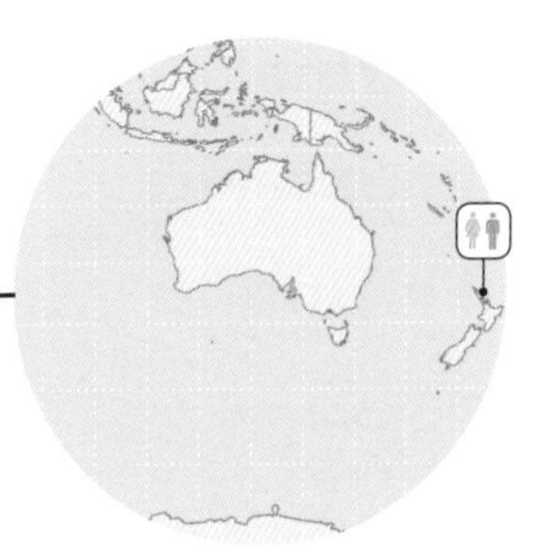

The reclusive expatriate, who lived in the North Island town for the last 25 years of his life, bequeathed Kawakawa with public toilets that have a live tree incorporated into the design, along with various integrated sculptures, wavy lines, irregular ceramic tiles and coloured glass.

Latrine, Enontekiö, Finland

This Arctic outhouse offers a pew with a view of Salmivaara Fell. It serves a wilderness hut at the west end of Lake Saarijärvi, on the Nordkalottleden Trail that wends through Enontekiö in Finnish Lapland. The trail, an epic 800km (500-mile) odyssey linking Finland, Norway and Sweden, is Europe's most northerly hike.

N 66° 20′ 13.2828′′ E 28° 22′ 50.0196′′

Toilet island, near Placencia, Belize

Eat your heart out Robinson Crusoe. This paradisiacal punctuation mark in the Caribbean Sea off Placencia, Belize, boasts its own flushing throne, from where the king or queen of the castaways can survey their desert-island domain. It's a long way to the shops when you run out of paper, though...

© CHRIS KOLACZAN / 500PX

Outhouse, British Columbia, Canada

Despite its ultra remote location on the shoreline of Haida Gwaii (formerly known as the Queen Charlotte Islands) in British Columbia, Canada, this impressive outhouse features an automatic flush, powered by the moon, which washes all waste away twice a day.

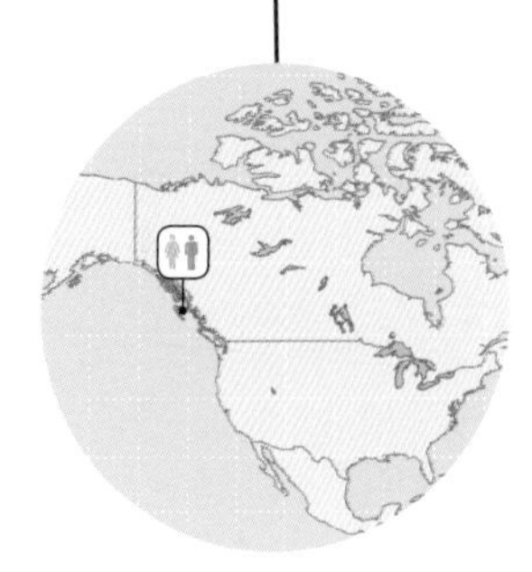

Eco-toilet, British Columbia, Canada

Yes, of course bears do… especially when the facilities are this swanky. Make like a grizzly and sit in the woods, on this uber green composting machine in Taylor Arm Provincial Park, a raw wilderness area on the north side of Sproat Lake in British Columbia, Canada.

'The Leaning Dunny of Silverton', Silverton, Australia

Australia is famous for its outback thunderboxes – where it pays to check under the seat for creepy crawlies before sitting down. The legendary leaning dunny of Silverton has become an iconic sit stop for people passing through the former gold-mining town, 25km (15.5 miles) from Broken Hill in regional New South Wales.

Latrine, Arnarstapi, Iceland

In Jules Verne's *A Journey to the Centre of the Earth*, Arnarstapi (or Stapi) in Western Iceland is the last place Professor Otto Lidenbrock, Axel and Hans stop before climbing nearby Snæfellsjökull and travelling into the bowels of the planet. Like all good last stops, it boasts an excellent toilet.

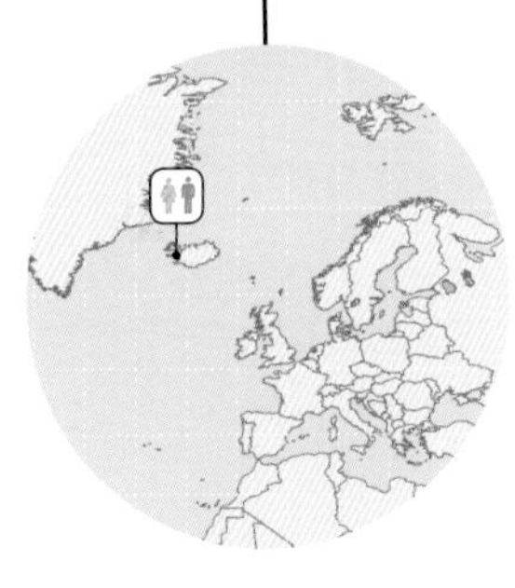

Mirrored restroom, Texas, USA

Like a scene ripped straight from a nightmare, this paranoia-inducing public restroom in Sulphur Springs, Texas, is made entirely of one-way mirrors, giving users the impression they're sitting in a glass box while the whole world walks past outside.

© CINDY ROLLER

In order to ensure the dignity of users is maintained, it needs to be lighter on the exterior than the interior, so at night, the shiny cubicle is lit with LEDs shone from outside the toilet, magnifying the feeling that you're on full display.

N 33° 46′ E 8° 23′

'Comfort toilets', Chott el Djerid, Tunisia

Chott el Djerid, a large salt lake in southern Tunisia, was used as the setting for Luke Skywalker's boyhood home in the original *Star Wars* film. The Lars' subterranean homestead may have been destroyed, but the Galactic Empire failed to extinguish the new hope represented by these roadside 'comfort' toilets.

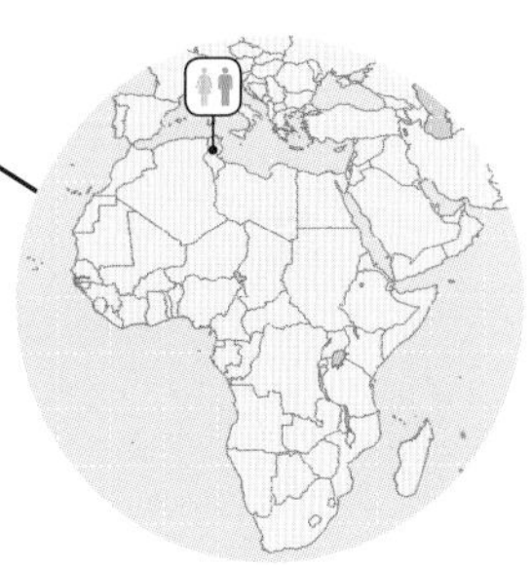

Safe Haven Orphanage, Ban Tha Song Yang, Thailand

A seemingly simple toilet block forms an unlikely cultural connection between a rural orphanage in Ban Tha Song Yang, Thailand, and a high-tech architectural firm from Trondheim, Norway. Conceived by 15 student architects engaged in a workshop held by TYIN, the Safe Haven sanitary station was built by local Karen people.

N 17° 33′ 19.8360′′ E 97° 55′ 17.4324′′

Eco-toilet, Encounter Bay, Australia

Tactfully painted to blend in with its bushland surrounds on the foreshore of wild Waitpinga Beach in Encounter Bay, South Australia, this eco-toilet serves a salty bunch of beach bums, who seek out the solitude, surf breaks and fishing spots offered by the Fleurieu Peninsula coastline.

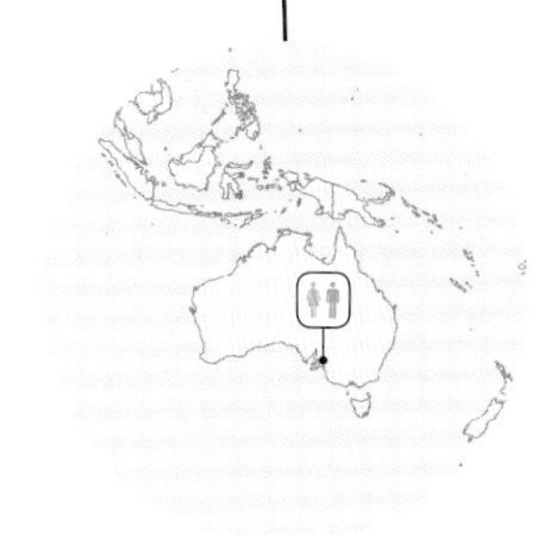

© VLADIMIR CUVALA / ALAMY STOCK PHOTO

Chata pod Rysmi, Slovakia

Stepping into the alpine air to visit the facilities at Chata pod Rysmi, a mountain chalet perched aloft in Slovakia's High Tatras, is hardly onerous when you're greeted by such views. The hut sits beneath Rysy, a triple-peaked mountain that straddles the border with Poland and climbs to 2500m (8200ft).

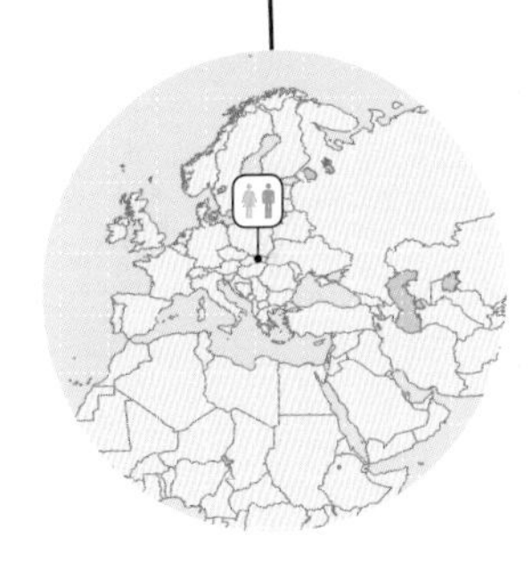

fotofoto Gallery, New York, USA

When you're exploring the fotofoto Gallery in Huntington, New York, the exhibits don't stop just because you've gone to the restroom, as seen here, with the 2013 American Family Gallery Project by artist Jeffrey R Smith. The curated images follow you right into the toilets too.

© JAMES CAPO / 500PX

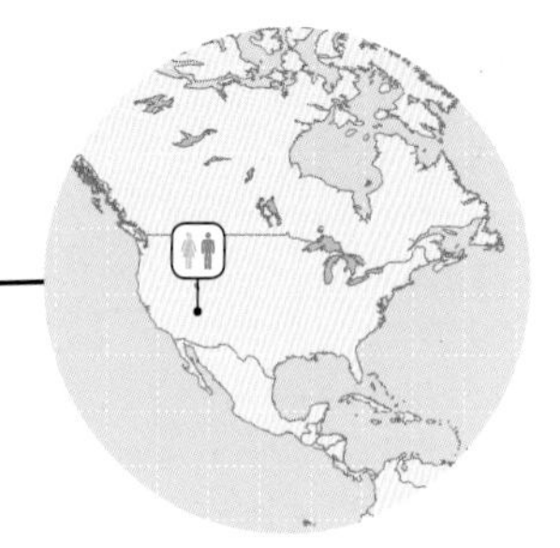

Tonto Trail, Grand Canyon National Park, USA

Limited privacy is the trade-off for sensational views from the hot seats of these composting campsite toilets on the 112km (70 mile)-long Tonto Trail through Grand Canyon National Park, Arizona. Instead of going rim-to-rim, the Tonto Trail traces the Colorado River, traversing the bench separating the inner gorge from the upper canyon.

Desert toilet, the Siloli, Bolivia

You can forget frills, privacy and shelter in this open-air latrine in the arid heart of Bolivia's Siloli Desert, but there's never a queue for the toilet. The Siloli, a continuation of the Atacama Desert in neighbouring Chile, is famed for wind-sculpted rock formations such as Arbol de Piedra (Stone Tree).

WC
BAÑO

Aoraki/Mt Cook National Park, New Zealand

A popular walk through stunning Aoraki/Mt Cook National Park – home to New Zealand's highest peak – is the Hooker Valley tramp, a 13km (8 mile) hike beside the eponymous river. This is wild country – that is, not a place where you'd expect to find a handily placed WC – but New Zealand is full of surprises.

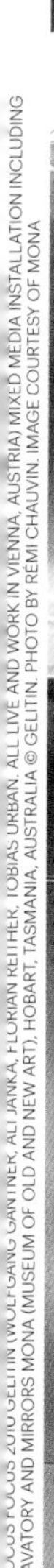

FOCUS FOCUS 2010 GELITIN (WOLFGANG GANTNER, ALI JANKA, FLORIAN REITHER, TOBIAS URBAN. ALL LIVE AND WORK IN VIENNA, AUSTRIA) MIXED MEDIA INSTALLATION INCLUDING LAVATORY AND MIRRORS MONA (MUSEUM OF OLD AND NEW ART), HOBART, TASMANIA, AUSTRALIA © GELITIN. PHOTO BY REMI CHAUVIN. IMAGE COURTESY OF MONA

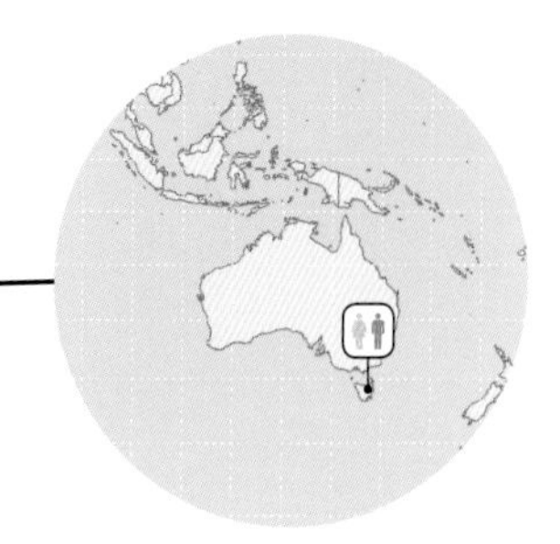

MONA's lavatory installation, Tasmania, Australia

Little is too taboo for Tasmania's MONA (Museum of Old and New Art) to display; the Australian gallery's permanent collection features the scatological concepts of the Viennese collective, Gelitin, who use mixed media (including the alarming combination of lavatory and mirrors) to create things you'll have trouble forgetting once seen.

Fjallabak Nature Reserve, Iceland

Within Fjallabak Nature Reserve in the Highlands of Iceland, on the jagged edge of Laugahraun lava field at the northern end of the popular 55km (34-mile) Laugarvegur hiking trail, is Landmannalaugar, which boasts geothermal hot springs, a mountain hut and this terrific triangular toilet with long-cooled lava lapping at its heels.

N 54° 18′ E 8° 39′

St Peter-Ording, Germany

The sea reigns supreme in Germany's Northern Friesland, with several villages having been entirely swallowed by storm tides in previous centuries. Around St Peter-Ording, the swell can reach 3m (10ft) high, and dunes have a habit of moving, so from 1910, the community began building structures on stilts – including the public toilets.

N 37° 56′ 29.9724′′ E 27° 20′ 28.0680′′E

Ephesus, Turkish Aegean

From the 1st century AD, the ancient city of Ephesus – an important Greco-Roman metropolis, located near present-day Selçuk in Turkey's İzmir Province – boasted modern-looking municipal toilets, complete with marble seats. In front of the 36 toilet holes in Scholastica Baths runs a trough where wiping sponges were rinsed after use.

© JAN PHILIPP KOHRS / 500PX

Waterfall washroom, Taroko National Park, Taiwan

The Baiyang Waterfall Trail in Taiwan's Taroko National Park might not be very long, but it boasts seven impressive tunnels and numerous curtain-style cascades along its 2km (1.2-mile) length. Appropriately, the washroom by the trailhead is fed directly by one of the path's waterfalls.

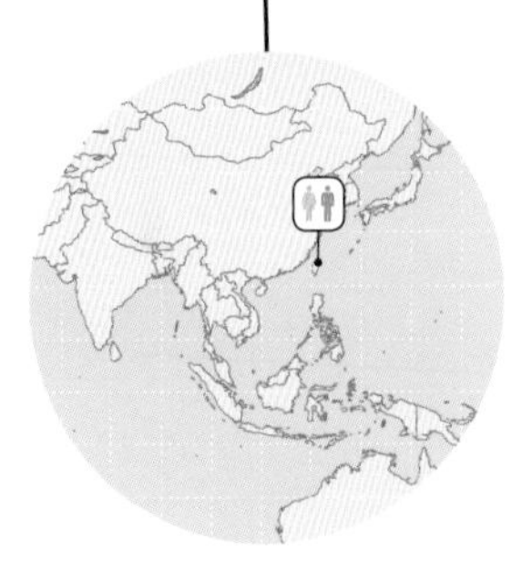

Public toilets, Calgary, Canada

Local authorities splashed out $250,000 on these arty farty twin toilets for Calgary's East Village in Alberta, Canada. The super high-tech, hipster-pleasing public conveniences automatically play music when the door is closed and locked, whereupon a sanitised toilet seat suddenly appears from the wall, clean and ready to use.

N 51° 2′ W 114° 3′

© PIERRE F / 500PX

Schönbrunn Castle toilets, Vienna, Austria

These leafy conveniences are found in the vast verdant grounds of Vienna's 17th-century Schönbrunn Castle. This 1441-room Baroque palace, complete with expansive gardens exquisitely manicured and shaped over successive centuries by the green hands of the Habsburg's royal gardeners, is one of the biggest attractions in the Austrian capital.

© YEHONATHAN ELOZORY / 500PX

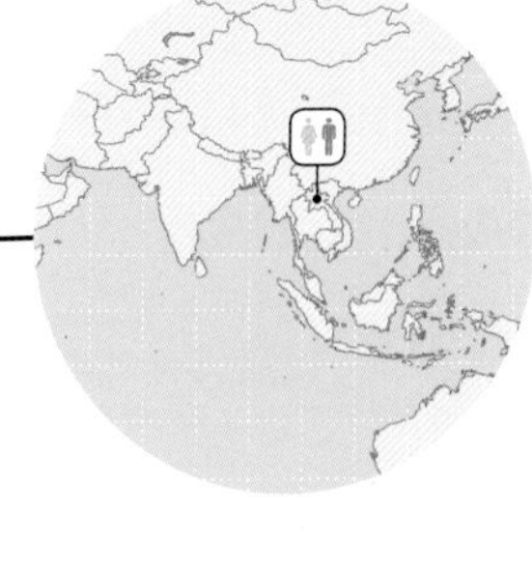

Valley view restroom, Laos

Sit, stand or squat – the choice is yours in this well-maintained roadside rest stop in the mountainous Southeast Asian nation of Laos. Whichever way you lean, the valley view from the loo, through a wide-open hole in the wall, is utterly uplifting.

N 43° 53′ W 69° 7′

Outhouse, Metinic Island, Maine, USA

A birder's nirvana, the lonely isle of Metinic sits off the coast of Knox County in Maine, in the chilly embrace of the North Atlantic. The island hosts a nesting population of Arctic terns and a handful of binocular-clasping humans, who can twitch while they queue for this outdoor toilet.

Fruity bathroom, Tel Aviv, Israel

Jaffa oranges are named after the oldest part of Tel Aviv, where the juicy fruit was first developed by Palestinian farmers in the 19th century. The sweet, virtually seedless citrus variety is now exported worldwide from Israel, which has honoured the orange with a series of spherical street toilets.

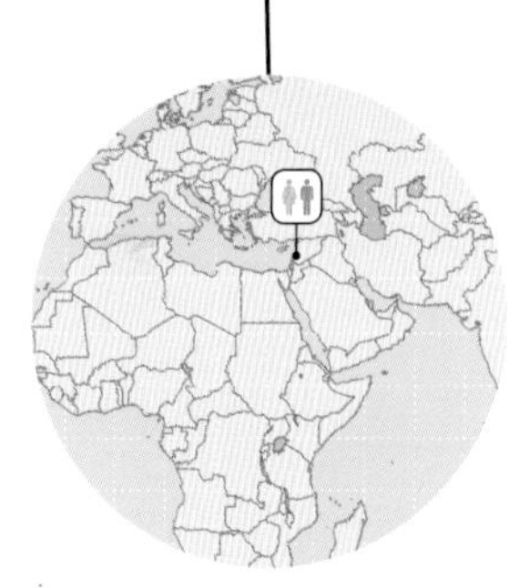

His ‘n’ hers, Jericoacoara Beach, Brazil

Since the *Washington Post* blabbed about Jericoacoara being one of the world’s best beaches, this erstwhile hidden gem on Brazil’s east coast has become a hotspot for travellers questing for blue lagoons, sun-blasted sand, tranquil seas and immense dunes. They’ve even had to build these his-and-hers palm-frond beach bogs.

S 2° 47′ W 40° 30′

Růže Hotel, Český Krumlov, Czech Republic

Sitting on the throne rarely feels quite as regal as it does in the boutique Růže Hotel in Český Krumlov. This wonderful wood-encased WC is found in rooms that once accommodated Jesuit monks during the 16th century, in the heart of the UNESCO World Heritage–listed Czech Republic town.

© ROELOF NIJHOLT / 500PX

Huldefossen waterfall, Norway

No need to run the tap while perching on this picturesque potty next to the cacophonous Huldefossen Waterfall near Førde in Norway; the sound of thousands of gallons of water rushing over the 90m (295ft) drop should drown out any unwanted acoustics. Norway boasts nine of the world's 20 highest waterfalls.

Red Woods toilets, Rotorua, New Zealand

Thanks to the geothermal activity around Rotorua in New Zealand, the 'Sulphur City' has a perpetual eggy odour; paradoxically, the public toilets in the Redwoods Forest are sweet as. The shrouds, designed by Maori artist Kereama Taepa, each depict a native North Island bird, which is either extinct or endangered.

Mr Toilet House, Suwon, South Korea

Suwon, in South Korea, boasts a theme park totally devoted to toilets. The eccentric attraction revolves around a commode-shaped museum, former home of Sim Jae-duck – aka 'Mr Toilet' – one-time mayor of Suwon and first president of the World Toilet Association, which strives to improve sanitation in developing countries.

© DAN SCHAUMANN / WWW.TOILOGRAPHY.COM

Jae-duck, who was born in his poor grandmother's outhouse, became famous after providing toilets for football fans during the 2002 World Cup. Sadly, he died in 2009 (not, as far as we know, in the loo) and never saw the results of his push to elevate the toilet to an *objet d'art*.

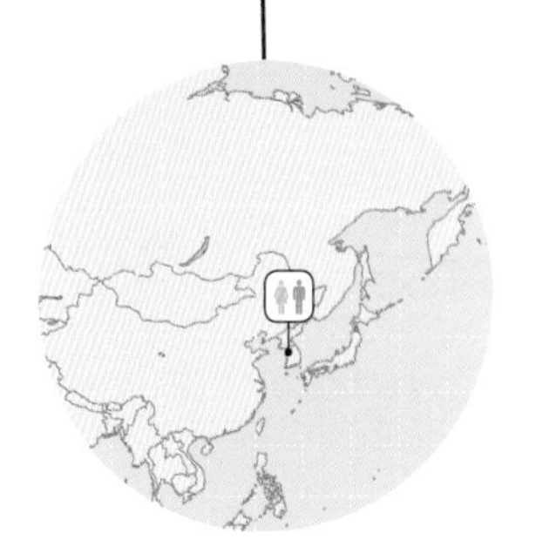

N 51° 29′ 56.4936″ W 0° 9′ 23.2560″

Ladies' boudoir, Mosimann's, London, UK

Housed in a former Presbyterian church, the Belfry in Belgravia, London, Swiss chef Anton Mosimann's club is one of the world's swankiest eating joints. Unsurprisingly, given the prices being paid and the expectations of the clientele, including HRH the Prince of Wales, the restrooms aren't half bad either – with the ladies' boudoir particularly sumptuous.

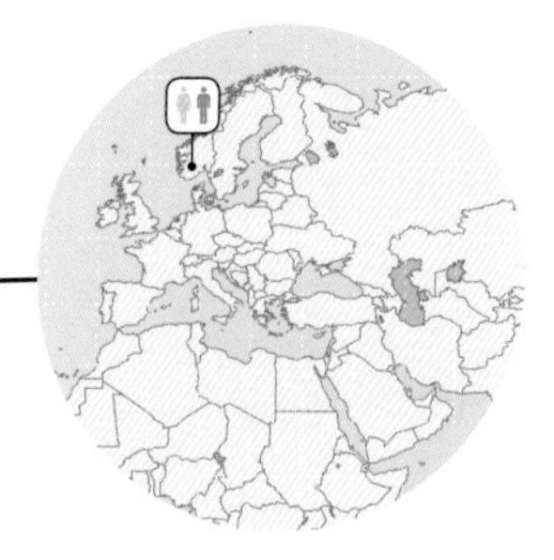

Snowy outhouse, Hjartdal, Norway

When it dumps in Norway, it really dumps – a snow storm has virtually turned this outhouse into a wood-lined igloo, as it stands contrasted against a crisp cobalt sky on a bluebird day in the mountains of Hjartdal municipality in central Norway's Telemark county.

Gents' toilets, UFO Bar, Bratislava, Slovakia

Before being reinvented as a restaurant, the flying-saucer-shaped building that now houses UFO in Bratislava, Slovakia was a Soviet-built observation tower, positioned to have a commanding view of the Danube. This vista can be enjoyed by everyone visiting the venue, but only gents can see it from this bucketlist angle.

© SHEYNE LUCOCK / 500PX

Gaudí-style toilets, Bahia, Brazil

Gaudí-style toilet blocks are a surprise addition to a sprawling Brazilian beach resort like Bahia's Praia do Forte, but they're just one of many modern additions that combine to confuse leatherback turtles that have nested here for thousands of years. Fortunately, a local sanctuary looks out for the bewildered beasts.

Segantini hut restroom, Switzerland

Austrian-born 19th-century painter Giovanni Segantini lived his last years in a St Moritz alpine aerie now known as Segantini Hut, capturing the Swiss peaks with his palette. The hut, perched at 2731m (8960ft), is currently a lodge, where visitors to the iconic outhouse enjoy eye-watering valley views of the Engadine.

Basham Beach Conservation Park, Australia

You can virtually see the sea from this eco-friendly unisex dunny in Basham Beach Conservation Park near Port Elliott on the Fleurieu Peninsula in South Australia.

If the waves were visible, you might even spot from here one of the migrating southern right whales that regularly cruise this coast during the winter.

© WERNER MONATSSPRUCH / 500PX

Sony Center, Berlin, Germany

Everything is uber modern in the Helmut Jahn–designed Sony Center in Berlin's Potsdamer Platz, including the men's room. The centre, completed in 2000, occupies a historic spot that was flattened during WWII, and became a no-man's land when the Berlin Wall was built across it during the Cold War.

N 65° 3′ 10.6704″ W 146° 3′ 23.7240″

Log outhouse, Chena Hot Springs Resort, Alaska, USA

If Santa has an outhouse, it surely resembles this log bog on the banks of a creek meandering through Chena Hot Springs Resort in Fairbanks, Alaska – though you'll have to be an employee to enjoy it. The resort also boasts an Ice Museum, featuring frozen carvings, including a life-size effigy of jousting knights and a depiction of a (non-functioning) ice toilet.

NOT A PUBLIC
OUTHOUSE
EMPLOYEES
ONLY

Lençóis Maranhenses National Park, Brazil

The dunes of Brazil's Lençóis Maranhenses National Park seem like an odd spot to encounter a lonely Tardis-like toilet block, but this place is full of mirage-like apparitions. Despite its arid facade, this 'desert' is just outside the Amazon Basin, and its sands are regularly punctuated by clear blue pools.

Art Deco restrooms, Miami Beach, Miami, USA

The Art Deco restrooms at Miami Beach break the surface like a submarine coming up for air. The South Beach – aka the American Riviera or Art Deco District – is so synonymous with the 20th-century architectural style that many buildings have been recognised by the National Register of Historic Places.

Haida Gwaii, British Columbia, Canada

This well-weathered and atmospheric outhouse stands amid a mossy green rainforest dreamscape at Skidegate on Haida Gwaii (formerly known as the Queen Charlotte Islands), in Canada's British Columbia. Skidegate is a Haida First Nations community, historically targeted by European settlers pursuing the lucrative trade in sea otter pelts.

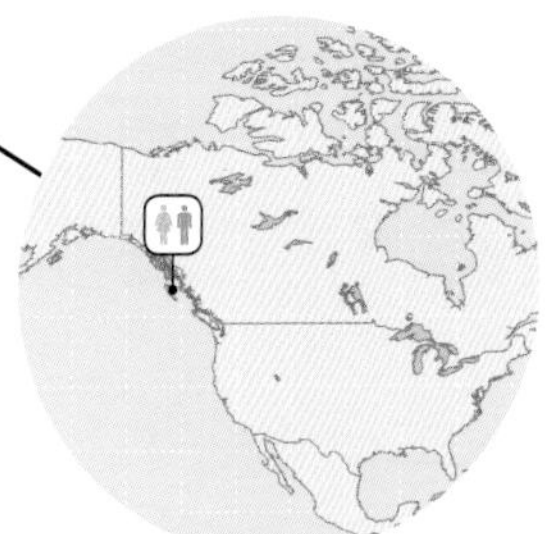

N 42° 41′ W 44° 30′

Outhouse, Kazbegi, Georgia

Balanced precariously on a wobbly wooden bridge above a fast-flowing snow-melt fuelled mountain river in Georgia, very close to the border with Russia, this rickety ensemble of Kazbegi outhouses enjoy the ultimate power flush, as provided directly from Mother Nature.

Ancient public lavatories, Carthage, Tunisia

Before its destruction by the Romans in 146 BC, Carthage was one of the world's most developed cities. At its peak, the capital of the Carthaginian Empire had 400,000 residents, multi-storey buildings, a sophisticated sewer system and dozens of public toilets, such as these, found close to modern-day Tunis.

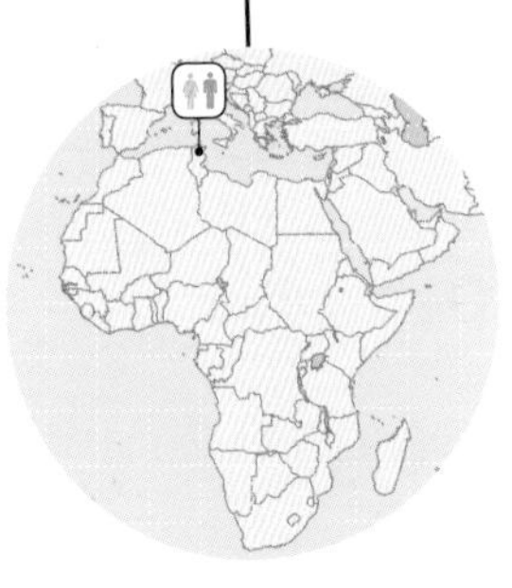

Roadside Reststop, Norway

When architects Manthey Kula were commissioned to replace a roadside restroom that had been blown away by Arctic elements in Norway's far north, they established two objectives: make it really heavy (so it won't blow away again) and build it without any windows at head height, so visitors get a break from the intensity of the landscape.

The Roadside Reststop Akkarvikodden is constructed from welded Corten steel, with the inner walls glass-screened to prevent rust rubbing off onto punters. The outside scene penetrates the design just once, in the smallest restroom, where a ceiling-mounted glass panel offers a reflection of the horizon.

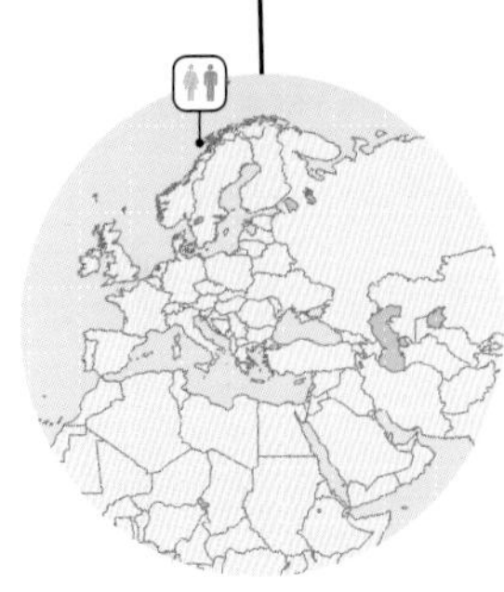

N 25° 47′ W 80° 8′

14th Street public toilets, Miami, USA

The Art Deco style that makes Miami Beach's Ocean Drive one of the world's most architecturally iconic roads, even extends to the toilets – such as this 14th St restroom, just along the street from Villa Casa Casuarina, former residence (and murder site) of Italian fashion designer Gianni Versace.

Milford Track, New Zealand

By the time you reach 1154m (3786ft)-high MacKinnon Pass, halfway around the Milford Track – one of New Zealand's Great Walks, oft-claimed to be the world's finest hike – you've truly earned a sit down. Watch out for keas – mountain parrots/flying pirates who will steal food from your backpack while you're otherwise engaged.

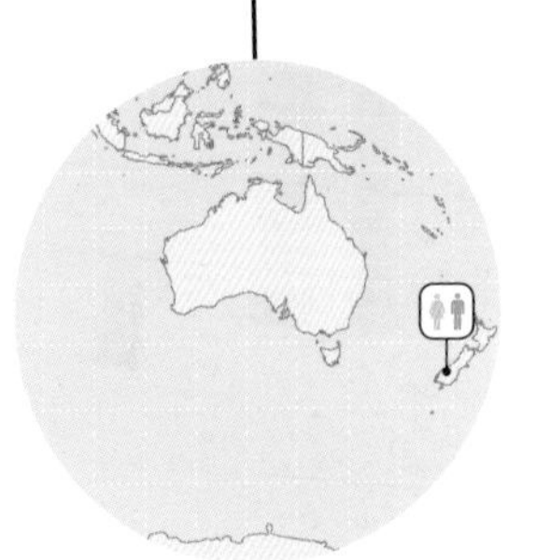

S 44° 49′ 45.4944″ E 167° 47′ 35.0664′

Milford Track, New Zealand

New Zealand's original Great Walk, the 53km (33-mile) waterfall-splattered, peak-punctuated Milford Track in Fiordland National Park on the South Island, was described by poet Blanche Baughan, writing in London's *Spectator* in 1908, as 'the finest walk in the world' – and she didn't even get to use these awesome outhouses.

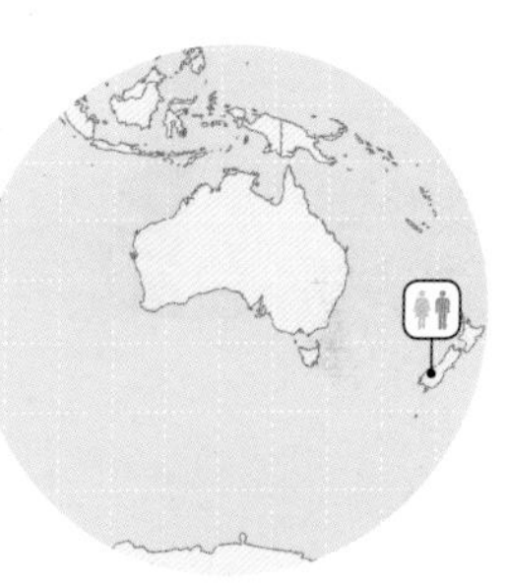

Quirimbas National Park, Mozambique

At Guludo barefoot beach lodge in northern Mozambique's Quirimbas National Park, visitors enjoy the ultimate long-drop loo with a view. After years of isolation during Mozambique's civil war, the park boasts blooming populations of land and marine animals, ranging from elephants, lions, leopards and crocodiles to dugong and sea turtles.

MICHAEL MELFORD / GETTY IMAGES ©

Yoho National Park, Canada

Hugging the western slope of the Continental Divide in the Canadian Rocky Mountains, Yoho National Park takes its name from the Cree Nation word for awe and wonder – words that will come in useful when describing the experience of visiting this wonderful outhouse with an awesome aspect on the park.

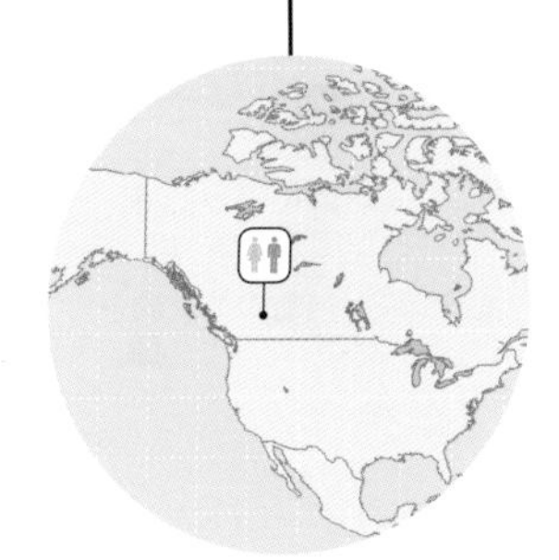

N 52° 28′ 18.5376″ E 13° 26′ 30.0228″

Café Achteck, Berlin, Germany

A classic Café Achteck. These green, 8-sided pissoirs blossomed around Berlin in the late 19th century to accommodate the bursting bladders of an influx of work-seeking men. Each featured seven urinals, with the eighth wall being the door. Of the 142 originals, those that survived wars and redevelopment are protected.

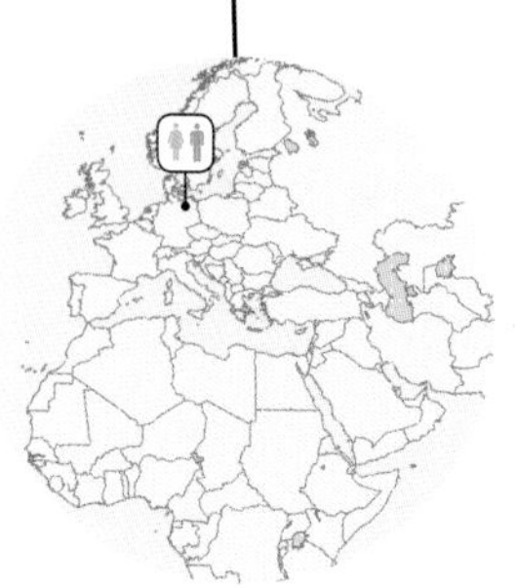

© OLAF MENZ / 500PX

Jonsknuten, Kongsberg, Norway

This restroom on the rubbly flanks of Jonsknuten in southern Norway looks up at the peak of the 904m (2966ft) mountain. You have to leave the door open to enjoy the view, but the chances of being interrupted by anyone bar an inquisitive troll are minimal.

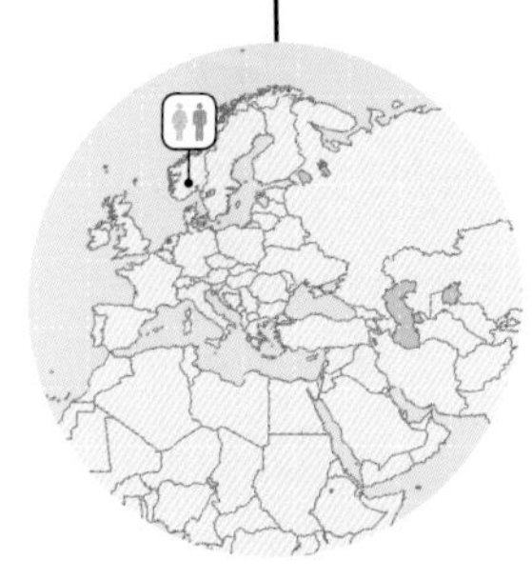

Toilets: a Spotter's Guide
April 2016
Published by Lonely Planet Global Ltd
ABN 36 005 607 983
www.lonelyplanet.com
3 4 5 6 7 8 9 10
Printed in China
ISBN 978 1 76034 066 7

Written by Patrick Kinsella

Managing Director, Publishing Piers Pickard
Associate Publisher & Commissioning Editor Robin Barton
Art Director Daniel Di Paolo
Layout Designer Lauren Egan
Thanks to Christina Webb, Jessica Cole, Karyn Noble, Jonathan Butler at Alamy, Laura Stanley and Lucas Luxton at 500px.com
Pre-Press Production Ryan Evans
Print Production Larissa Frost, Nigel Longuet

COVER IMAGES: © ROBERT DOWNIE, © DAN SCHAUMANN/ WWW.TOILOGRAPHY.COM, © LINDA MCKIE / GETTY IMAGES, © COLINGRAINGER / 500PX, © "ZUMA PRESS, INC. / ALAMY, © MICHAEL MELFORD / GETTY IMAGES, © CHRIS PINCHBECK/GETTY IMAGES, © MEINZAHN / GETTY IMAGES

Lonely Planet Offices

AUSTRALIA
The Malt Store, Level 3, 551 Swanston St, Carlton, Victoria 3053
T: 03 8379 8000

IRELAND
Digital Depot, Roe Lane (off Thomas St),
Digital Hub, Dublin 8, D08 TCV4

USA
124 Linden St, Oakland, CA 94607 T: 510 250 6400

UK
240 Blackfriars Rd, London SE1 8NW T: 020 3771 5100

STAY IN TOUCH lonelyplanet.com/contact

Paper in this book is certified against the Forest Stewardship Council™ standards. FSC™ promotes environmentally responsible, socially beneficial and economically viable management of the world's forests.